Cléber Alessandro Corso
Márcio L. Kalkmann
Ivete Linn Ruppenthal

Economic analysis on a small rural property

Cléber Alessandro Corso
Márcio L. Kalkmann
Ivete Linn Ruppenthal

Economic analysis on a small rural property

An analysis for decision-making

ScienciaScripts

Imprint

Cover image: www.ingimage.com

This book is a translation from the original published under ISBN 978-613-9-62500-0.

Publisher:
Sciencia Scripts
is a trademark of
Dodo Books Indian Ocean Ltd. and OmniScriptum S.R.L publishing group

120 High Road, East Finchley, London, N2 9ED, United Kingdom
Str. Armeneasca 28/1, office 1, Chisinau MD-2012, Republic of Moldova, Europe
Printed at: see last page
ISBN: 978-620-7-74576-0

SUMMARY

DEDICATORY..2
ACKNOWLEDGMENTS..3
EPISODE..4
SUMMARY..5
1 INTRODUCTION..6
2 LITERATURE review..8
3 METHODOLOGY..24
4 PRESENTATION AND ANALYSIS OF RESULTS..28
5 FINAL CONSIDERATIONS..49
6 REFERENCES..51
APPENDIX A - QUESTIONNAIRE APPLIED TO THE OWNER..56

DEDICATORY

To my parents Henrique and Geneci, for life and for the principles they passed on to me.

To my friends for the moments of joy and sadness we experienced, because they were the ones who gave me the strength and courage to keep going, things that no amount of money can pay for.

To God, for the simple fact that I can give thanks every day for every miracle that happens in my life.

THANKS

First of all, I would like to thank my family for giving me the support I needed to complete this course. To my mother for all her dedication to me, often giving up doing things for herself to devote herself to me, even without me asking. My love for you is priceless.

To my father for being my greatest example of victory in this life and for having taught me all the ethical and moral values I know. I would like to say that being with you in this life is a great privilege and you make me proud every day as a great man, a great father.

To my advisors Màrcio Leandro Kalkmann and Ivete Linn Ruppenthal for their help, commitment and dedication to the development of this study.

To my great friend, colleague and brother Anderlei Rodrigo Mahl for giving me total support and encouragement in difficult times, and to my other friends who are in my life, making it greater every day.

EPISODE

"If I've come this far, it's because I've stood on the shoulders of giants. "

Isaac Newton

SUMMARY

The market is becoming more challenging every day due to major changes in politics, the economy and culture. In turn, technology is becoming increasingly advanced, changing the profile of society. This requires the adoption of strategies. Strategy is about how organizations conduct themselves in the context of competition and market positioning. An organization with a good strategy has an advantage over others, because its objectives are aligned and its managers have an idea of how to carry out their actions with extreme completeness, to which end they use management tools to ensure that they can achieve their estimated results. Rural properties, regardless of their size, also need decision-making strategies. Given this context, this study maps and evaluates three sources of resources on a particular rural property, with the aim of describing business management on small rural properties, emphasizing the need for an ideal evaluation of the viability of economic activities, as a strategic management tool, bringing data, statements and results achieved, in order to bring a better perception of the agricultural segment to rural properties and bringing data analyzed through mathematical models that can maximize the profits of the property under study. The research is characterized as exploratory-descriptive, and is classified in two aspects: in terms of its aims, it is of an applied nature. As for the means, the research was bibliographical, documentary and a case study. Data was collected by means of bibliographical research, documents and interviews, for subsequent analysis and discussion of the results. It was necessary to obtain the costs of the activities, as well as total production, turnover and market prices, in order to have a basis for calculations to analyze and describe the opportunity costs of the three activities on the property, in order to decide where to direct the property's investments. Thus, through this study it can be identified that under current conditions, dairy production is unfeasible, bringing an optimal profit maximization model that would be the joint production of corn and soybeans, thus maximizing profit.

Keywords: Rural property. Analysis. Viability.

1 INTRODUCTION

The current world scenario portrays the need for constant change and innovation within the most diverse forms of organization. The search for stability is characterized by constant improvement within companies. Such considerations are present in a rural organization, where agricultural *commodities* are traded, which have the same or more significant characteristics of seasonality in market prices. Rural properties are constantly being developed, given the need for continued food production in the world, as announced by the media, government bodies and experts on the subject.

Over the years, there have been countless discussions about the choices to be made regarding strategies to generate wealth, reduce inequality and improve the living conditions of people in rural areas. A path considered modern is one that frames management activities within scientific norms, aiming for more immediate economic results, but with long-term vigor, bringing producers closer to consumers. Choosing this path has shaped a belief based on the principles of agricultural economics that has predominated in regional agricultural modernization policies.

Analyzing a business is an important management tool because it is one of the most useful things a businessman or entrepreneur can do and have for making decisions based on this tool. It can serve both as a guide to be followed and as an instrument used to obtain funding to constantly improve the property. Planning allows mistakes to be confined to paper, rather than made on the market. It tests the decision to open or expand a business, since preparing a comprehensive business analysis requires time and effort, as well as knowledge of the market in which the business is located.

This work is part of the discussion on rural development based on the diversity of family farming, the mechanization of the activity and the creation of product portfolios, as well as strategic business units. The aim of this study is therefore to carry out an economic analysis of a small family farm.

According to Pinazza & Araùjo (1993, p.112), "in developing countries, investments in infrastructure in rural areas are justified, regardless of whether or not rapid urbanization occurs, because they leverage economic growth and development".

Family farms generally feel the need to work in the best possible way in their productive environment, optimizing the environment, reducing costs and improving productivity with the resources available on the property. In this way, we can see that there is a need to professionalize the management of processes on this property, thus justifying this study.

Santos, *et. all.* (2002) point out that a rural property can become an excellent source of income if its management is as effective as any successful business. The essence of economics means that, through the use of its tools, it has a technical view of the business, regardless of its purpose. This vision provides guidance so that all decision-making is aligned with the objectives of the business managers.

A rural property needs efficient management in order to manage, predict risks and plan investments. In view of this need for professionalization within a production system, this study aims to answer the following question: How can production be optimized with the resources available and control over the activities carried out through an economic analysis on a small rural property in the municipality of Tucunduva?

In this sense, this study focused on a few specific points, since the subject is very broad. The objectives of the study are presented below.

The general objective of this work is to study and characterize the tools used to manage economic activity on a rural property. The specific objectives of this study are:

a) Map the factors of production and detail the economic activities currently carried out on the property under study;

b) Carry out a survey of the expenses and income of the property studied;

c) Identify the opportunity costs between alternative activities;

d) Evaluate the activities that maximize and enhance the economic results of the property studied.

This study is structured in chapters. Chapter 2 presents the theoretical framework, which covers subjects related to the topic in order to deepen knowledge. Chapter 3 presents the methodology, which describes the steps required to carry out the study.

Chapter 4 presents the analysis and discussion of the results obtained from the research and, finally, the final considerations, which show that the objectives proposed in the study have been achieved, the problem proposed in the study is answered, the research problem is answered and future work is suggested.

2 LITERATURE REVIEW

Efficient management is the fundamental pillar for any project to be able to develop well in any business format. Entrepreneurs in rural businesses face a number of difficulties when it comes to managing an enterprise or company in this sector. The first step is to start an economic analysis, identifying the main points that can be considered, the most significant challenges or problems for management, in order to bring positive results that will add to the sustainability of the business. The following subjects will be covered: Business Strategy; Economic Approach and Profit Maximization; Agribusiness; Rural Activity Costs, Opportunity Cost, Management Models and Business Plan.

2.1 BUSINESS MODEL AND BUSINESS MANAGEMENT

Nowadays, companies are in a process of constantly reviewing their business models, organization and management. The analysis of methods and concepts that may have worn out, as well as the re-evaluation of procedures, are activities that are indispensable to the operating and productive mode that current practices demand, and which need to be applied at different levels within an organizational structure, so these models can be applied to rural companies, where a business model adds value to the current scenario of the properties (ROSA, 2013).

2.1.1 A brief approach to economics and profit maximization

Profit is considered a residue of the value of sales, after the different factors of production have been remunerated. In equilibrium, there is no remuneration left for the entrepreneur, except that which he earns as the owner of some factor among those combined in production (DANTAS *et al.,* 2002). The entrepreneur is identified only as the owner of these resources and is remunerated in this capacity.

Knowing that a decision is made to increase sales to the detriment of profits, there must be a minimum level of profitability, which represents a guarantee of financial security for the company. For each level of profit rate, there will be a maximum growth rate, corresponding to the retention rate, which ensures the flow of dividends compatible with the lower limit of security in the valuation of shares by the market (POSSAS, 1985).

According to Stoffel (2013), technological change, productivity, profit maximization, advanced or backward producers, are terms used by authors with a neoclassical approach when dealing with aspects of production, which, despite being microeconomic, are analyzed in macroeconomic terms. The increase in productivity as a way of reducing costs and improving income, the acquisition of modern inputs and the expansion of production are some of the defenses of the neoclassical authors. In this theoretical context, agriculture is defined as modern, traditional or backward SCHULTZ (*apud* STOFFEL, 2013). The focus is on satisfying the demands of the market, and it is up to the producer to offer products at adequate prices, for which there is demand and which guarantee the maximization of profits, with the least interference from the state.

The neoclassical approach became more present from the 1970s onwards, in the case of Brazilian authors. Regarding Brazilian agriculture, Buainain, Romeiro and Guanziroli (2003, p. 313) mention that "in the 1970s it was believed that the 'agricultural question' had been overcome by the modernization process, based on mechanization and the use of selected varieties of seeds and chemical inputs". This view, defended by some of the experts of the time, who believed in the supremacy of large producers, due to the scale of production they were able to achieve, also argued that these larger producers would remain in the countryside, while those with small tracts of land (or small producers) would be relegated to the rural exodus (STOFFEL, 2013).

Marshall (*apud* Stoffel 2013) argues that the main defense is economies of scale, according to which companies that manage to operate at larger sizes achieve productivity gains and, consequently, lower costs. Even though it wasn't a theory developed for agriculture, it is common for larger production scales to be advocated as a way of increasing competitiveness.

Nor does Schumpeter (1988) value innovation as an important factor in improving the use of available resources and obtaining higher profits. What is needed is the ability to use the resources you have in an innovative way, increasing the company's profitability. The result of innovation is savings, but access to credit and the distribution of property and income are important for this. For the author, innovation is creative and leads to development, which is why the innovative entrepreneur makes all the difference. Their sole objective would be to maximize profit, achieved through constant modernization of the means, both technical and organizational, at their disposal (STOFFEL 2013).

2.1.2 Strategy

Strategy is a bridge between the means and the objectives to be achieved. Quinn (2001, p. 136) defines strategy as "the pattern or plan that integrates the main goals, policies and sequences of actions of an organization into a coherent whole". However, he also says that the objective of strategy is to order and organize the resources available for viable and recurring action, based on the organization's strengths and weaknesses in the face of environmental changes and the action of intelligent competitors.

According to Hambrick (1980), strategy is a multidimensional concept, making it difficult to define the word correctly. According to Vicari (2013), in the organizational field, the term strategy began to be used more intensively soon after the end of the Second World War, as competition became fiercer during this period. Chandler *apud* Vicari (2013) said that "strategy is the determination of a company's basic long-term objectives, the adoption of appropriate actions and the obtaining of resources to achieve these objectives".

Along the same lines, Porter (1980) defines the concept of competitive strategies as offensive or defensive actions that are used to create a defensible position in an industry or business, to successfully face competitive forces and thus obtain a higher return on the expected investment. The diversity of perspectives in the study of business strategies has contributed to the concept of strategies appearing in the specialist literature with multiple meanings, which are not always properly clarified and which correspond to particular ways of approaching the issue and operationalizing the concept. Thus, the author analyzes some of the meanings to which the concept of strategies is generally linked (VICARI 2003).

For Slack (2002), strategy is the general pattern of decisions and actions that position the organization in its environment and that aim to achieve its long-term goals. A strategy has both content and process. The content of a strategy concerns the specific decisions that are made to achieve specific objectives. The process of a strategy is the procedure that is used within a company to formulate its strategy.

Strategies must be present in the management of small farms because rural activity today is considered to be a business and farmers are the entrepreneurial managers of the business, who face the same challenges as normal companies as a whole. This makes the role of the farmer a major factor in deciding on the strategies to be taken to best position their products on the market (VICARI, 2003).

Every organization, regardless of its line of business, its area of operation or its size, adopts objectives to be achieved directly or indirectly. The manager's main objective is always to maximize profits while optimizing costs, but everything depends on the direction and coordination of efforts and the organizational structure in order to survive in an increasingly competitive market (QUINN, 2001).

2.1.3 Strategies in the Context of Rural Properties

Rural properties are part of an environment made up of other rural properties, companies that supply inputs and technology, companies that buy production, research groups and other companies and entities that are part of the rural environment. Sette (1999) warns that managers in this process must conduct their business within this dynamic of interaction, in other words, conduct their business in accordance with how the market is reacting to constant changes, taking into account that all changes can occur within this environment.

According to Sette (1999), planning rural properties is an extremely important stage. During planning, the most frequently asked questions by farmers are:

- What to produce?
- What activities are best suited to the company?
- Which are the most profitable?
- What is the ideal combination of these activities on the property?
- What is the property's purpose?

According to the same author, for the business to succeed it is essential to understand the context in which the producers are working, i.e. what the potential and limits of the available infrastructure are, and to identify which agents interfere in agricultural production and how they act.

For Prahalad & Hamel (1995), the quality of a business will no longer be a competitive differentiator, but the main requirement for entering the market. As well as improving in this respect, companies will have to reinvent their sector. Strategies will always have to be renewed, because production processes, customers, are basically the criteria for promoting managers and measures for evaluating business success. It will be necessary to make a new adjustment to the sector in which the company operates. This shows that companies are not just competing within the boundaries of existing sectors, but are working to create the structure of future sectors. There is a need for transformation, both on the organizational side and in the sector as a whole. A consistent structure is needed to compete in a constantly changing market.

2.1.4 Family-run rural establishments

Family farming is understood as one in which the family, while owning the means of production, takes on the work on the productive establishment. So, a family farmer is any farmer whose main source of income is agriculture and whose workforce is based on family members, and Todesco (1999) says that it is important to stress that this family character is not just a superficial and descriptive detail, in other words, the fact that a productive structure combines family-production-work has fundamental consequences for the way it acts socially. Just like a company, it needs to look for resources and a market to sell its produce, but it is of paramount importance to the production chain, with a major influence on food production around the world.

Agriculture, seen as a business, is leaving behind its survivalist attitude, as stated by Santos, Marion and Segatti (2002, p. 13): "... this is the moment of transition that demands changes in the management of their businesses, mainly leaving their traditional posture of farmer for that of rural entrepreneur". For Passos *et al.* (2006, p. 25), the concept of family consists of a "social group united by biological, legal and/or affinity ties, which is constantly changing. It has its own criteria for inclusion, qualification, recognition and evaluation".

Rural companies can be characterized by a number of points, according to (Oliveira, 1999 *apud* Braum, 2011), including: strong emotional ties which, as a rule, influence behavior and decision making, difficulties in separating the rational from the emotional, and an attitude of austerity, both in terms of investment and in managing the company's expenses, but with an expectation of high loyalty, where political skill is often more valuable than administrative ability.

Braun (2011) emphasizes the three-circle model, which has been widely accepted in the business world. It is shown in figure 1.

Figure 1 - The Three Circles of Family Business model

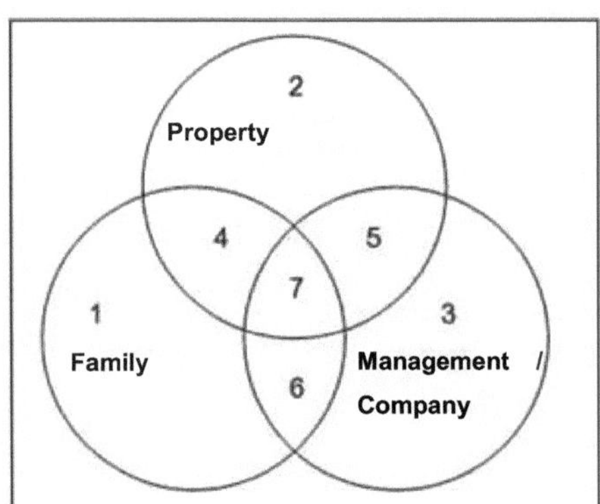

Source: GERSIK et al 2006.

According to Brandt *et al.* (2015), Braun (2011) characterizes each sector of the family business as follows:

a) Sector 1: is occupied by a family member who is neither an owner nor an employee;

b) Sector 2: is occupied by a shareholder who is neither a family member nor an employee;

c) Sector 3: is occupied by an individual who is active in the organization, but is not a shareholder and therefore has no ownership in the enterprise;

d) Sector 4: is occupied by an owner. He is also a member of the family, but not an employee;

e) Sector 5 is represented by the owner who works in the company but is not a member of the family;

f) Sector 6 represents the individual who participates in the management of the organization and is a member of the family, but is not the owner;

g) Sector 7 is represented by the individual who is both the owner, a member of the family and participates in the management of the organization.

This model by Gersik *et al* (2006) is present as a concept in countless articles and books, thus demonstrating the veracity of the subject in question and its importance in the evolution of family businesses.

Santos *et al.* (2002) also state that:

> Rural managers should be aware that the greater their knowledge of the structure and functioning of the unit and the factors of production, the greater the chances of improving their economic results (SANTOS *et al.* 2002, p. 19).

The same author states that among the knowledge about the structure of the organization, some can be identified which, when known by the rural manager, increase the reliability of decisions. Such knowledge would be:

a) Farm size - Measured by production capacity, profitability and return on investment, not by number of animals and planting area alone.

b) Yield of crops and livestock - This is the amount of production obtained by a unit of production factor. To achieve this, you need to work with good quality crops, plant them with good seeds and look after them properly.

c) Selection and combination of productive activities - This factor represents the choice of crops and livestock to be grown and the area that each should occupy on the property. It is important to know the whole area of the property very well, making a map of the areas of each type of soil. This enables us to use the land rationally, conserving the soil and achieving good yields. "In addition, we should make an inventory of all the property's assets, with the aim of making full use of them in productive activities."

Santos *et al* (2002) states that the combination of activities is therefore the amount of land that each crop, livestock and breeding will occupy on the property. The best combination is the one that makes the most intensive use of all the factors of production that exist in the company and allows them to be well conserved, so that fixed costs are lower per unit produced and profits are consequently higher.

The 2006 Agricultural Census identified 4,367,902 family farming establishments, which represents 84.4% of Brazilian establishments. This number of family farmers occupies an area of 80.25 million hectares, or 24.3% of the area occupied by Brazilian agricultural establishments. These results show a still concentrated agrarian structure in the country: non-family establishments, despite representing 15.6% of all establishments, occupy 75.7% of the occupied area. The average area of family establishments is 18.37 hectares, and that of non-family establishments is 309.18 hectares. This shows the great productive potential of family-run rural establishments (IBGE, 2006).

The 2006 Census carried out by the Brazilian Institute of Geography and Statistics presents significant figures related to Family Farming in the country: Of the approximately 5.1 million agricultural establishments in the country, more than 4.3 million are characterized as family farmers, representing 84% of the total (IBGE, 2006).

Figure 2 - Number of agricultural establishments (in millions)

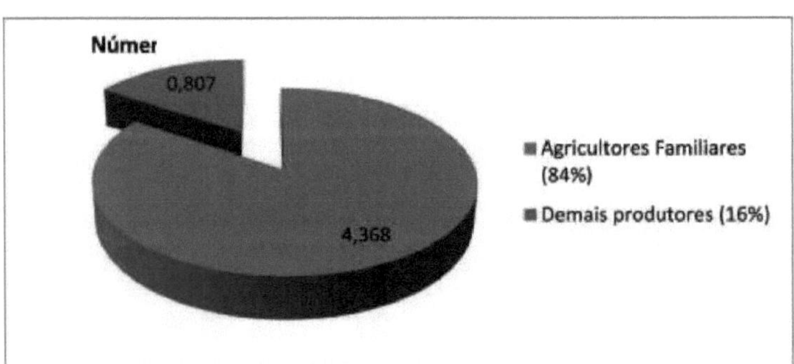

Source: CoDAF, 2016

Of the 16.5 million people who carry out some kind of rural activity, 12.3 million are related in some way to Family Farming, making up 74% of the total.

Figure 3 - Number of people working in the countryside (in millions)

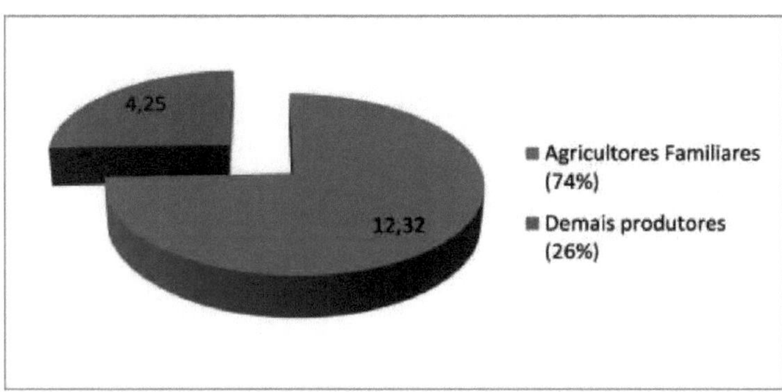

Source: CoDAF, 2016

Of the R$143.3 billion generated by the national agricultural sector, R$54.3 billion comes from Family Farming, accounting for 38% of the total;

Figure 4 - Amount of value produced by the agricultural sector (in billions)

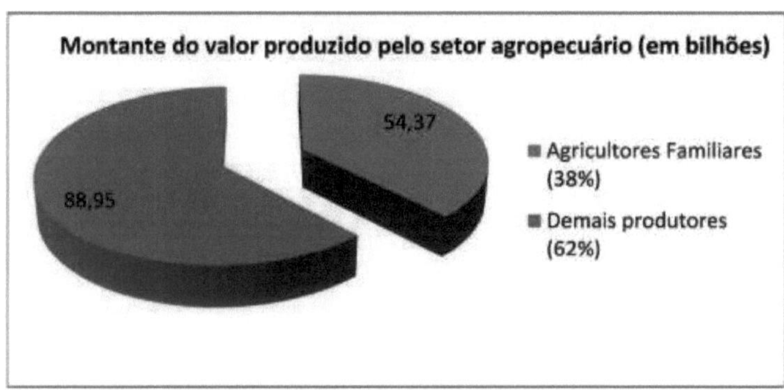

Source: CoDAF, 2016

The area occupied by family farmers corresponds to 80.2 million hectares, which represents 24.3% of the total land used by agricultural establishments in the country.

Figura 5 - Area occupied by agricultural establishments (in million/ha)

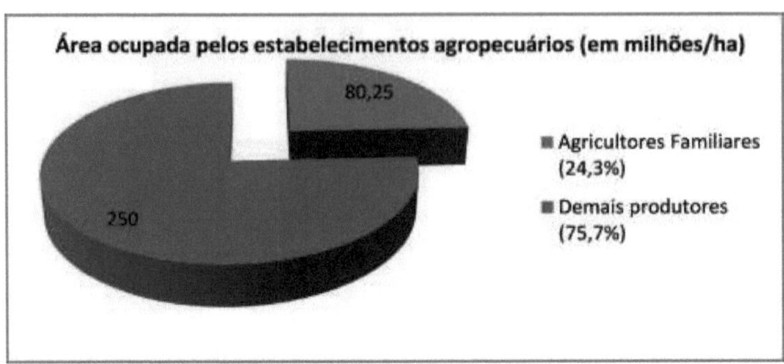

Source: CoDAF, 2016

This shows how important family farming is in the country's economy, even if it doesn't have the visibility that large-scale production has, especially when it comes to exports. The work carried out within family enterprises guarantees domestic supply in line with the population's food demands, creating an environment conducive to reducing hunger and local development, income and well-being in the countryside.

Furthermore, the 2006 Agricultural Census reveals that more than 80% of rural producers are illiterate or have not completed high school, the vast majority of them are illiterate or can read and write, but have no schooling (39%) or have incomplete primary education (43%), totaling more than 80% of rural producers. Among women, who account for around 13% of agricultural establishments, illiteracy reaches 45.7%, while among men the rate is 38.1%. The rates for other levels of education are: 8% for complete primary education, 7% for complete agricultural technician or secondary education, and only 3% with higher education.

Figura 6 - Graph of the distribution of producers in establishments, by level of education, according to the Major Regions-2006

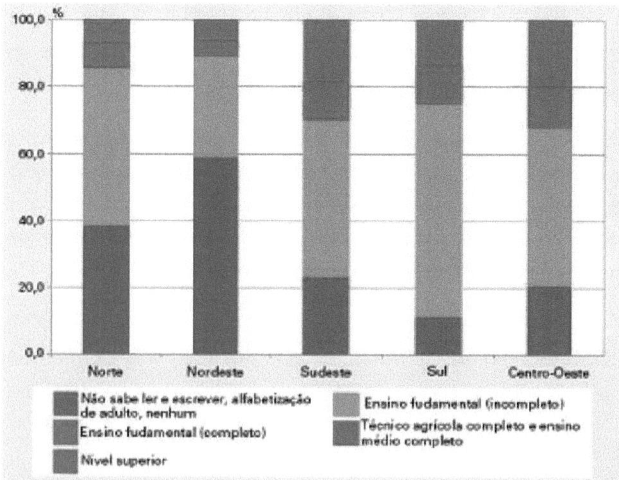

Source: Agricultural Census, 2006 (*apud* IBGE, 2016).

This shows that the level of schooling contributes significantly to the inefficiency of the management of rural properties, demonstrating that rural organizations increasingly need incentives ranging from education to professional incentives, because the market is constantly changing, and they need to adapt to these changes so that they can continue in the market, in order to maintain the livelihood of the property.

2.2 RURAL ACTIVITY

According to Padovane (2006), accounting is the social science that aims to record and control the economic, financial and administrative acts and facts of entities, while also studying assets. It is an information and evaluation system designed to provide its users with economic, financial, physical and productivity statements and analyses in relation to the entity being accounted for. There are various accounting concepts for the many segments of this science, such as: cost, commercial, industrial, banking, etc. The following are definitions of rural accounting.

According to Calderelli (1997, p. 180), "Rural accounting is accounting based on the orientation, control and recording of acts and facts that occur and are practiced by a company whose object of commerce or industry is agriculture or livestock". Marion (2003, p. 29) comments that "individuals considered to be large rural producers will be treated in the same way as legal entities for accounting purposes, and must keep regular accounts, through a qualified accounting professional".

According to Oliveira (2008, p. 27), "the purpose of accounting is to analyze, interpret and record the phenomena that occur in the assets of physical and legal persons", and this tool is necessary for all individuals who use it, through reports or useful information to be able to make the right business decisions. However, Oliveira (2008) states that:

> Thus, the producer of the rural property (individual or legal entity), by means of the information generated by accounting, is able not only to control costs and evaluate results, but above all to establish plans and draw up strategies that take the property towards production efficiency, always observing the specificities of each type of crop, the market and the best technology (OLIVEIRA, 2008, p. 27).

According to Crepaldi (2006), rural accounting is an important ally for rural entrepreneurs, as it provides clear information to help them make decisions. It is one of the main information control systems for rural companies.

According to Gomes (2002, p. 21):

> Rural accounting is a fundamental instrument for the financial and economic control of rural property; it can also be said that the use of accounting contributes, in many ways, to the environment in which the entity operates.

In order to control production costs and expenses, producers need to use management control. The technological development of information shows rural producers the possibilities of systems for adopting administrative procedures and making decisions, allowing them to improve their financial and economic results (VELOSO; FERNANDES; BARIONI, 2003).

According to Crepaldi (2005), Rural Accounting is a tool that is little used by rural producers, as it is seen as a complex technique, with a low return in practice, and is only known for the Income Tax Return, so producers show no interest in its managerial application. Among other factors, he pointed out that what has contributed to this is the deficiency of accounting systems, which are responsible for portraying the characteristics of agricultural activity, as well as the lack of professionals trained in transmitting administrative technologies to rural producers, hence the failure to include Rural Accounting as an instrument of government agricultural or tax policies.

Rios (2008, p. 13) points out that "every well-advised organization performs better. In a rural entity, this premise is also true. If rural owners made more use of rural accounting tools, they could achieve better results".

2.2.1 Opportunity costs

In the economic approach, human beings guide their decisions based on the premise of optimization, supported by the assumptions of objective rationality and people's freedom of action. In other words, if they are free to act, it is logical to assume that they try to choose things that provide them with maximum satisfaction. Miller (1981, p.4) refers to opportunity cost as a "model of rational behavior, where the alternatives for action in a decision are evaluated in a systematic and coherent way, and the choice of the best option is bounded by the limitations of the real world".

Therefore, for economic theory, the opportunity cost arises when an individual decides on a particular alternative direction at the expense of other viable and exclusive alternatives. As such, it represents the benefit lost by choosing a particular alternative over others. In this way, the cost of factors of production can only be measured through their opportunity cost. Miller (1981, p.188) points out that "cost has a very special meaning in economics, it means only one thing - the opportunity cost. "

> [...] the cost of factors for a company is equal to the values of these same factors in their best alternative uses. This is the doctrine of alternative or opportunity costs, and it is what economists accept when they talk about production costs (BILAS, 1980, p. 168).

According to Varian (1994, p. 352) "the economic definition of profits requires us to evaluate all inputs and outputs at their opportunity costs. "

In order to better clarify the concept of opportunity cost from an economic point of view, the table below shows some definitions of the term expressed by various economists:

Chart 1- Concepts of opportunity cost from an economic perspective

MEYERS (1942, p.194)	The cost of production of any unit of merchandise is the value of the factors of production employed in obtaining that unit - which is measured by the best alternative use that could have been made of the factors if that unit had not been produced.
BILAS (1967, p.168)	The cost of factors for a company is equal to the values of these same factors in their best alternative uses.
LIPSEY & STEINER (1969, p.215)	The cost of using something in a specific endeavor is the benefit sacrificed (or opportunity cost) by not using it in its best alternative use.
LEFTWICH (1970, p.123)	The cost of a unit of any resource used by a firm is its value in its best alternative use.

Spenser and Spielgelman, (*apud* MAURO, 1991, p 170), suggest that opportunity cost refers to:

> (...) the cost of the opportunities given up, or in other words, a comparison between the policy that was chosen and the policy that was abandoned. For example, the cost of using capital is the interest that can be earned on the next best use with equal risk. If capital funds can earn 5 percent on their most productive use, then that is their cost to the company using the funds.

For Stiglitz (2003), opportunity costs are the true investment costs of the inputs used. According to Varian (2006), when making an economic definition of costs, it is necessary to evaluate each factor of production in terms of its market price, i.e. its opportunity cost. Pindyck and Rubinfeld (2009) distinguish between accounting profit and economic profit. For the author, accounting profit is the difference between revenue and cash flows related to the payment of labor and raw materials and interest and depreciation expenses, while economic cost also takes into account the opportunity cost. Therefore, the economic cost is higher than the accounting cost.

According to Cecconello and Ajzental (2008), Opportunity Cost represents the cost associated with a given choice measured in terms of the best opportunity lost. In other words, the opportunity cost represents the value that economic agents place on the best alternative that they forego when making their choice.

According to him, the opportunity cost should be considered when choosing an economic activity in agriculture, as it is a key factor in decision-making, so that the best decisions can be made for the progress and prospecting of the business, so that it flows upwards, always aiming for the best result for the activity.

For Saulit (1997), knowing what your direct and indirect costs are, or opportunity costs, is of paramount importance for the progress and development of agribusiness, thus bringing more sustainability and security to the business world. Opportunities in the sector are increasing every day, responsible for increasing market efficiency through marketing and management techniques, carrying out procedures that assist and plan the control and organization of the managerial activities of rural activity.

2.3 AGRIBUSINESS

Within any country, economic activity is divided into three sectors: primary, secondary and tertiary. The primary sector is made up of products that are sent to other industries to be transformed into industrialized products and which use a large amount of labor and land. The secondary sector is made up of activities that process and/or combine primary products, in which the capital factor is heavily used. The tertiary sector is defined as the set of activities that provide services, either for itself or for the other sectors of the economy (BACHA, 2007).

Today, agriculture can no longer be separated from the other sectors of the economy, which are responsible for all the activities that guarantee the production, processing, distribution and consumption of food. According to Mendes (2007), this systemic view of the agents that generate production is called *agrobusiness, agribusiness* or the agro-industrial complex. These groups of activities as a whole often have a *commodity* nature, since in their primary or semi-processed form they are very similar and can be classified according to exact specification standards. In this case, economies of scale are an important source of cost reduction and competitive advantage.

According to Mendes (2007), agribusiness goes beyond the boundaries of "rural property" to involve all the other sectors that participate directly or indirectly in the process of bringing food to consumers. In other words, agribusiness doesn't just encompass the individuals who work directly in the activity, but all the other factors that provide inputs, those who process the products that come from the activity, the individuals who manufacture the food, transport it and sell it to the end consumer, in other words, agribusiness isn't just the rural activity itself, but a chain that involves a large number of participants until the product reaches its end consumer. Still in Mendes' view, "all these operations are part of chains that have become increasingly complex as agriculture has modernized and the agricultural product has started to add more and more services that are outside the farm" (MENDES, 2007, p. 48).

According to Araùjo (2005), the "agriculture" of yesteryear, or the primary sector, is now dependent on many services, machines and inputs that come from outside.

It also depends on what happens after production, such as warehouses, various infrastructures (roads, ports and others), agro-industries, wholesale and retail markets, exports.

Based on Mendes (2007), the agribusiness system can be represented as follows:

Figure 7 - Visualization of the agribusiness system

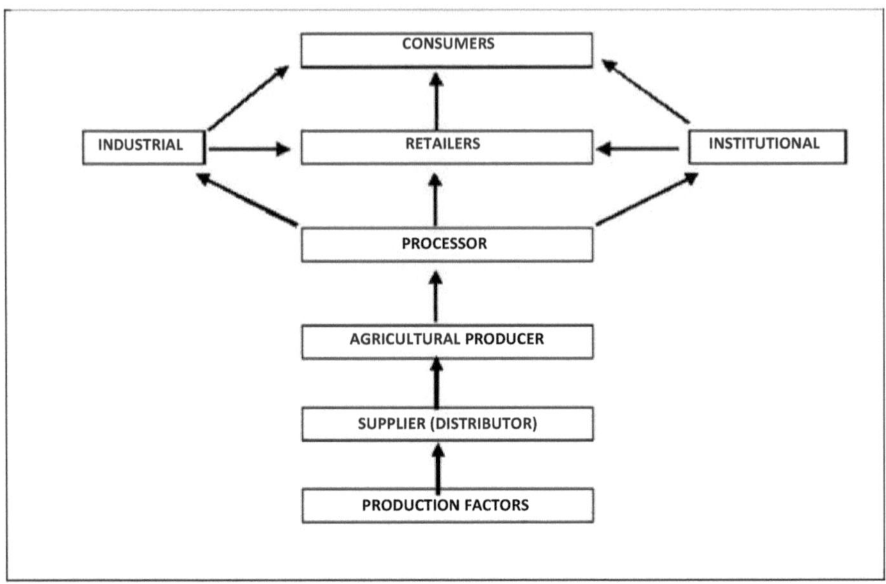

Source: MENDES (2007)

According to Rufino (1999, p. 17):

> This systemic view of the agricultural business - and its consequent treatment as a whole - offers great potential benefits for a more intense and harmonious development of Brazilian society. To this end, there are problems and challenges to overcome. Among these is knowledge of the interrelationships between production chains, in order to identify the requirements for improving their competitiveness, sustainability and equity.

According to Araùjo (2005, p. 15), "rural properties are increasingly losing their self-sufficiency, needing infrastructures, conquering markets, facing globalization and the internationalization of the economy".

According to Guilhoto *et al* (2000), around 70% of agricultural production is used in other sectors of the economy, with agribusiness and marketing absorbing approximately 72% of this total. These are the fastest growing segments of agribusiness, while the relative share of agriculture and industry in agriculture tends to decline. The result is that the agro-industry and commercialization segments are the ones that concentrate the most income in agribusiness.

According to Contini (2001), agribusiness is of paramount importance in generating income and wealth for the country. In terms of the agriculture variable, it is the economic sector that occupies the most labor, with 17 million people, together with 10 million of the other members of agribusiness, representing 27 million people in total. This sector occupies the most labor in relation to the value of production: for every R$1 million, the number of people employed in 1995 was 182 in agriculture, 25 in mineral extraction and 38 in construction.

In today's agribusiness scenario, success depends, among other things, on planning, management and the manager's technical and administrative skills in making the most of available resources, such as land, machinery, implements, human resources and the property's infrastructure, as well as information on internal and external production factors, The producer must therefore use management tools to organize and show the income, costs and expenses accounts, the economic scenario and other variables in the decision-making process (REIS, *et al,* 2011).

3 METHODOLOGY

In terms of its nature, this study is applied research, with immediate objectives that generate products or processes. As such, it used knowledge from basic research to solve problems related to concrete applications. According to Barros and Lehfeld (2000, p. 59), applied research is "motivated by the need to produce knowledge for the application of its results, with the aim of contributing to practical and didactic purposes, aiming at a short or medium-term solution to the problem encountered in reality", in other words, a useful tool for finding solutions.

With regard to the objectives, exploratory-descriptive research was used because, as Gil (2005, p. 35) explains, "exploratory research aims to explain and provide a better understanding of the research problem. In this type of research, the researcher seeks greater knowledge about the subject under study". According to Gil,

> A study aims to describe the facts and phenomena of a given reality or a given case study with the intention of getting to know a given subject, such as its characteristics, values and related problems, i.e. descriptive research has as its main purpose the description of the characteristics of a given population or phenomenon, or the establishment of relationships between variables (GIL, 2005, p. 96).

This study is classified as exploratory and descriptive because it aimed to describe the property and how it is structured. After collecting the data, an analysis was carried out that served as the basis for structuring the business plan, thus exploring the revenue and cost data that the property has, in order to have a basis for the plan.

In terms of approach, this study was classified as quantitative-qualitative. Qualitative research is concerned with gathering data on a particular subject, understanding and interpreting the behavior, opinions and expectations of individuals about something to be studied. This research has the environment as an open source of data and the researcher as its main instrument, because in this type of research the researcher comes into direct contact with the study environment with all the possible variables to be examined (GIL, 2002).

The qualitative research aimed to analyze and describe the management of small rural properties, thus meeting the first objective.

The second objective is to analyze the factors of production and the property's economic activities, while the fourth objective is to collect data on the property's income and expenditure, in order to put the analysis into numbers, so as to have a better understanding.

A quantitative approach is characterized by the use of quantification, i.e. numbers that explain a certain reality of the case studied, both in the information and in the treatment of variables through statistical techniques that aim to explain the facts obtained through numbers. The qualitative approach is a task of "organizing, summarizing, characterizing and interpreting the numerical data collected. To this end, the data can be processed using statistical methods and techniques" (MARTINS and THEÓPHILO, 2009, p.107).

This approach was used to quantify the costs, revenues and quantities produced by each activity carried out on the property, bringing a mathematical model through the *Microsoft Exel* tool to analyze it, also bringing it in the form of graphs that gave a better understanding of the analysis of the results obtained.

As for the technical procedures, documentary, bibliographic and case study methods were used. Documentary research is similar to bibliographical research, but the difference lies in the nature of the sources: while bibliographical research is based on authors, documentary research is based on materials that have no authorial or scientific definition (GIL, 2002).

According to Lakatos and Marconi (2001), documentary research is the collection of data from primary sources, such as written documents belonging to public archives, private archives of institutions or households, or statistical sources. Bibliographical research, on the other hand, is a survey of bibliography already published in books, magazines and articles of a scientific nature, and its purpose is to put the researcher in direct contact with everything that has been written on a given subject. It aims to provide the researcher with reinforcement in the analysis of their research or in the correct manipulation of information. For the same authors, bibliographical research:

> [...] it covers all bibliography that has already been made public in relation to the subject under study, from single publications, bulletins, newspapers, magazines, books, research, monographs, theses, cartographic materials, etc. [...] and its purpose is to put the researcher in direct contact with everything that has been written, said or filmed on a given subject [...] (LAKATOS and MARCONI, 2001, p. 183).

Therefore, the bibliographic research fulfilled the objective of describing management in rural properties, business strategies, especially in small properties, emphasizing economic analysis as a strategic management tool. While the documentary research fulfilled the objective of characterizing the business management of the rural property under study, where a survey of revenues, costs, profits (surpluses), expenses (investments) and indebtedness of the economic activity on the property studied was carried out, for which these data were collected from invoices and notes from the producer.

The case study was used as a strategy to delve deeper into the "how" or "why" or "how much" of a specific subject of study. Thus, the case study is neither a tactic for collecting data, nor merely a feature of the planning itself, but a comprehensive research strategy (YIN, 2005). Still based on technical procedures, a case study, according to Gil (2002, p. 128), "consists of the in-depth and exhaustive study of one or a few objectives, in such a way as to allow a broad and detailed knowledge of them". Its purpose is to explore real-life situations whose boundaries are not clearly defined; to describe a situation in the context in which the research is being carried out.

> Case study research faces a technically unique situation in which there will be many more variables of interest than data points, and as a result relies on multiple sources of evidence, with the data needing to converge in a triangle format, and as another result benefits from the prior development of theoretical propositions to drive data collection and analysis (YIN, 2001, p. 33-34).

Therefore, the case study fulfilled the objective of mapping the factors of production and detailing the economic activities currently carried out on the property, data which is specific to the property and does not apply to other properties due to their particularities.

In this way, this research is characterized as: applied in nature, in relation to the objectives as exploratory-descriptive research, with a quantitative-qualitative approach, using documentary, bibliographic and case study procedures.

To collect the data, a structured interview was carried out with the owner in order to gather information on the management of the property, which followed a script as shown in Appendix A, where the interviewer was given wide and unlimited access to the documents and information, and the interviewee was asked to answer the questions on several occasions.

Data was also collected by researching the documents provided by the owner, so that the information obtained could be analyzed in order to prepare an economic analysis of the business pertinent to the case studied, The values expressed in the analysis of the study have been corrected to 2014 prices, using the deflator based on the IPCA (consumer price index) To calculate the accumulation of inflation as well as other rates we must use the following mathematical formula:

$$i_{acumulada} = \left[\left(1+\frac{i_1}{100}\right)\times\left(1+\frac{i_2}{100}\right)\times\ldots\times\left(1+\frac{i_n}{100}\right)-1\right]\times 100$$

Where "ïΓ,"ï2","in" represent the rates that will be accumulated in their percentage value, divided by 100 to obtain the unit interest rate, so that we can add them to 1 (one), multiply them together, subtract the result of adding them together by 1 (one) and multiply again by 100 to obtain the accumulated percentage value.

The property is located in the interior of the municipality of Tucunduva-RS, chosen by the author of this research due to its ease of access and convenience. After collecting the data, the following tools were used to help analyze the data: *Microsoft Excel* to prepare the cost and revenue spreadsheets, as well as modeling the graphs, a mathematical model (solver), and linear programming to help with the analysis and preparation of the fourth objective.

4 PRESENTATION AND ANALYSIS OF RESULTS

This chapter contains the economic characteristics of the property under study, as well as the presentation and analysis of the results obtained through a questionnaire applied to a rural property, the analysis of this data obtained by verifying which activity maximizes profits for the property, as well as an economic analysis of the business that can guide the owner in which activity he should focus his efforts in order to obtain the greatest return.

4.1 CHARACTERIZATION OF THE PROPERTY

The rural property studied in this work is located in Vila Sâo Miguel, in the interior of the municipality of Tucunduva-RS. It is classified as a smallholding, as it has an area of between one and four fiscal modules. This classification is defined by Law 8.629, of February 25, 1993, which takes into account the fiscal module that varies according to each municipality in the national federation.

According to the National Institute for Colonization and Agrarian Reform (Incra), a fiscal module in the municipality of Tucunduva comprises 20 hectares. The rural property under study currently has a total of 25 hectares, which is 1.25 fiscal modules. Of these 25 hectares, 3 are set aside as legal reserves, 17 hectares are used to grow corn and soybeans and 5 hectares are used exclusively for dairy farming.

The property works with three sources of income, namely milk production, corn and soybeans; part of the corn production is used for dairy production and part is sold on the market, as is the soybeans, which are sold entirely on the market.

Table 2 below shows the gross revenue from all the activities carried out on the property. The data refers to the crop years 13/14, 14/15 and 15/16, with dairy farming referring to the calendar years 2014, 2015 and 2016. The values analyzed, calculated and deflated according to the IPCA (Consumer Price Index) are based on 2014, where in 2015 accumulated inflation reached a level of 10.67% and in 2016 it reached 6.29%, so the accumulated in 2016 reached 19.96% in relation to the base year of 2014.

Table 2- The property's gross revenue in years

Safra	Activity	Total Production	Average Price	Gross Sales
2014	Dairy	83,438 Ltrs	R$ 0,81	R$ 67.496,00
13/14	Corn	1507.55 Sc/60 kg	R$ 26,40	R$ 39.799,32
13/14	Soy	317.47 Sc/60 kg	R$ 66,08	R$ 20.978,20

Total				R$ 128.273,52
2015	Dairy	87,837 Ltrs	R$ 0,81	R$ 71.734,26
14/15	Corn	1486.23 Sc/60 kg	R$ 23,00	R$ 34.183,37
14/15	Soy	441.7 Sc/60 kg	R$ 60,68	R$ 26.802,36
Total				**R$ 132.719,99**
2016	Dairy	80,876 Ltrs	R$ 1,05	R$ 84.976,94
15/16	Corn	1929.42 Sc/60 kg	R$ 34,70	R$ 66.950,76
15/16	Soy	712.5 Sc/60 kg	R$ 70,76	R$ 50.416,50
Total				**R$ 202.344,20**
Total invoiced in the period				**R$ 463.337,71**

Source: Corso, Kalkmann and Ruppenthal, 2017.

Table 2 shows the total annual production of each activity, as well as the average market price and total gross revenue for the property over the period studied. It can be seen that in nominal terms, in 2014 and 2015 dairy farming had the same average market price, of approximately 30% in the average marketed value of the product, in terms of production there was an increase of 5% from 2014 to 2015 and a decrease of around 8% in production for the year 2016, while in terms of turnover from 2014 to 2015 there was an increase of 6.27% and from 2015 to 2016 there was an increase of 18% in the gross invoiced value. Bringing the figures into real terms, deflating them, it can be seen that the average price actually fell by 33.41% in the amount invoiced, and it can also be seen that although the average price of milk practiced on the market did not vary from one year to the next, but discounting inflation, the real value fell by 11.11%, i.e. the sale value in 2014 was R$0.81 and bringing it to real values in 2015 the sale value was R$0.72, impacting on a fall in real turnover of 5.06%.

Figure 8- Graph of gross sales per year/harvest at market prices

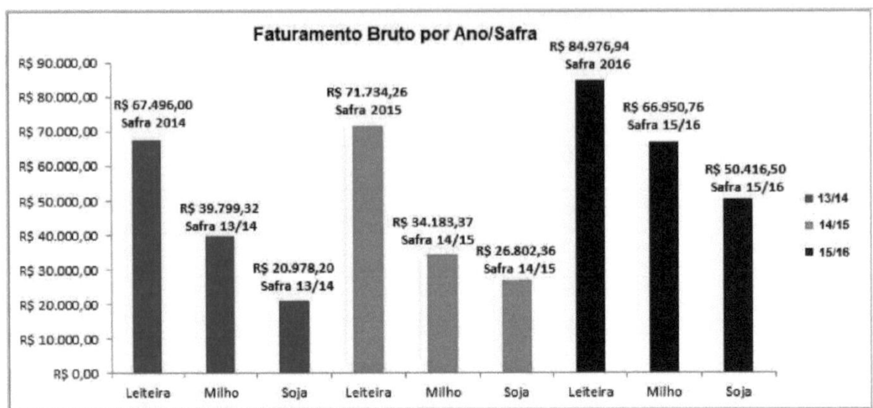

Fonte: Corso, Kalkmann e Ruppenthal 2017.

As shown in figure 8, in relation to the corn crop there was a 2% drop in the quantity produced, as well as a 13% drop in the average value practiced during the year, culminating in a 15% decrease in the total value invoiced from the 13/14 harvest to the 14/15 harvest, while based on the 15/16 harvest there was a 30% increase in production, due to the increase of 1 hectare of planted area, However, the average nominal value practiced on the market increased by approximately 51% in relation to the previous harvest, so the turnover was 95% higher than the turnover obtained in the previous harvest. In real terms, there was an increase of only 68.22% in turnover and the price practiced on the market was 25.90% higher than in 2014.

In the soybean crop observed on the property, there was an increase in production as well as an increase in gross revenue, analyzing between the 2013/2014 and 2014/2015 harvests, there was a 39% increase in total production, also with a 61% increase in nominal gross revenue, taking into account the real values practiced in the market, this increase in revenue was obtained in 27,76%, the average price practiced at real values fell by 25.48% compared to 2014, while in the 2015/2016 harvest there was an 88% increase in turnover and a 27% increase in production, all due to an increase of 3 hectares in the area planted, i.e. a 30% increase in the area planted.

When analyzing and comparing the years based on the 2013/2014 harvest, there is a big difference between the real and nominal values, as can be seen in the revenue per harvest. Nominally, the 2015/2016 harvest had an increase of 140.33% in its revenue compared to the 2013/2014 harvest, However, when compared to real market values, this increase was only 42.88%, as well as the comparison made with the corn crop, which nominally had an increase in turnover of 68.22% when in reality this increase was only

1.06%, having a major impact on the property's figures. An interesting case can be seen in dairy production, where the nominal turnover was 25.90%, but when it is brought to real values, the property saw a 36.73% drop in turnover, all due to inflation during the period studied.

The relative data for milk production is based on the calendar year, i.e. everything that was produced and invoiced in the period from 01/01 to 31/12 of each year in the period studied. For corn and soybean crops, the agricultural year is considered, which refers to the months of May to April, counting together the invoicing and production in the planted areas, including harvest and off-season.

4.2 STRATEGIC PROPERTY MANAGEMENT

A questionnaire was administered to the owner in order to gather some data about the property and to characterize the current management and how the property's activities are progressing.

When asked if he had ever had any kind of training or had any reference to what a business plan is, he reported that this variable had been discussed several times in informal conversations with technicians from supplier companies and Emater technicians, but it had never been put into practice, i.e. this aspect had never been discussed in more depth, which he believes is one of the main factors in the smooth running of a property.

When it comes to managing farm costs, he points out that the sources of information in this process are mainly the empirical knowledge acquired and the factors analyzed when defining the activity/crop to be developed.

At the moment, the owner mentions that the only technology used on the property is the equipment used for milking, and that no technological or IT devices are used to help control the property.

Regarding the question of the correct quantities of inputs in the activities carried out on the property, he says that the correct amount is used, but he has no basis to follow, no technical support to help him with the necessary combinations of inputs used in production.

When asked if there are any possibilities for reducing costs, the owner says that there are always possibilities for reducing costs. The methods used would be to carry out a broader market research, nowadays only two suppliers are used due to the convenience and ease of access to inputs, it would be necessary to carry out a more in-depth price research, and also to reallocate activities, a more in-depth analysis of the factors of production on the property would be interesting in order to be able to allocate activities in a better way.

When asked if he does market research, he said that he doesn't use market research because he works with only a few suppliers for whom he buys products over a long period of time, which means that he gets some benefit from the purchase. Because no price research is done, the purchase of inputs is according to the availability of delivery from the suppliers in question, since the purchase is not made with stock, but rather on the need for consumption of the same.

Regarding the threats and opportunities of the activities carried out on the property, he says that: The threats are the uncertainties of market prices, because the property works with *commodities* (cereals and fat cattle) which prices are not formed in the local market, and milk which also has its price defined by the local market, so it is hostage to the price variation of buyers. The climate is also an important and decisive factor in the production and development of the property, in this case it is a direct factor that defines a rise or fall in production, both in the production of grain and milk.

The opportunities are that he can focus on which branch or which crop he will work on most intensively, so he can assess which will bring him the greatest profitability.

The owner has been in the sector for more than 40 years. What has improved is that in the past it wasn't so easy to produce and sell, because you could only sell to suppliers/buyers and you were held hostage to the market, and technology in handling has come on strong to help things run smoothly and make it easier to carry out activities, and it has also brought about a reduction in the labor required to carry out activities.

Technological innovation has had a major impact on the property, because at the beginning, the work was all manual labor, requiring a workforce of five people or more to carry out all the activities on the property. Today, with just two people working, all the activities can be carried out quickly and practically.

At the moment, the owner has no interest in changing activities, but rather in focusing on a particular activity in the production range so that it brings a higher return. However, the owner has no knowledge of what would be the ideal choice of activity to achieve such a return.

4.3 MAPPING THE ACTIVITIES CARRIED OUT ON THE PROPERTY

Mapping the activities carried out aims to describe all the functions and details, from costs to income on the property, its production, its characteristics, in order to provide a broad view of the activity for analysis of the results presented.

4.3.1 Mapping Dairy Production

The property underwent a period of changes to its structure in the years prior to the data being collected, with the aim of adapting the processes to a better way of carrying out the activities, from improving the facilities and equipment to changing the structure from a "manual" way of working to a semi-manual way of working, which has reduced labor and made it easier to carry out the activity more practically. The property works exclusively with Jersey breed animals because they are smaller and produce a higher quality product.

The property has been working in this field for more than 30 years. At the beginning, they worked with a small number of animals, and the milk produced was used to make cheese. Over the years, the activity has become stronger on the property, and they have seen an income opportunity with a greater financial turnover. During the period studied, the property had an average of 18.42 lactating cows with an average production of 12.68 liters/cow per day.

The following data, shown in figure 9, represents the milk production delivered to Cotrimaio (Cooperativa Agropecuâria Alto Uruguai Ltda.) from January 2014 to December 2016.

Figura 9 - Graph of total milk production delivered during the period

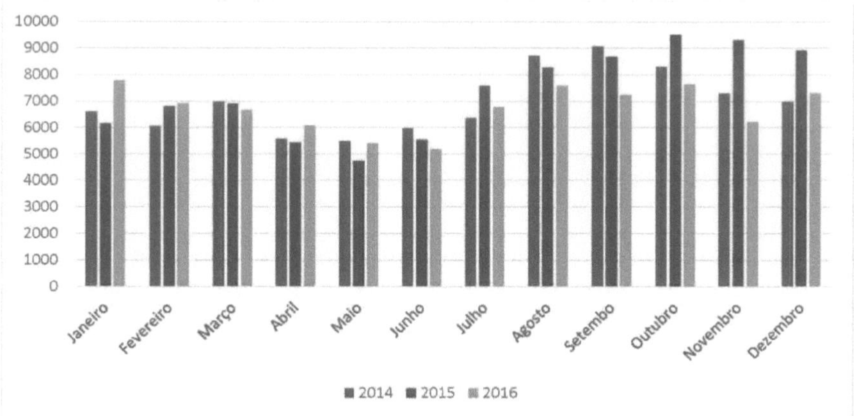

Source: Corso, Kalkmann and Ruppenthal, 2017.

This graph represents production at an average number of 18.4 lactating cows over the total period studied.

In order to interpret milk production, several factors need to be taken into account, such as: the stage of lactation, the age at which the cow is producing, the animal's physical condition, dry matter consumption, the types and nutritional values of the food consumed by the animal, the climate, etc. All these factors influence the amount of milk produced per animal, which will consequently affect the final turnover.

It can be seen that in the months of May to June production falls, due to the scarcity of green feed, with more use being made of feed and silage, which reduces production and consequently increases costs. From July to December there is an increase in production due to the planting of winter pastures and then the planting of summer pastures begins. The peak in production is at the end of winter and the beginning of spring, in the months of August to October, when the animals give birth, i.e. new milk animals, which produce a greater quantity in their first 4 months, increasing production. The months of December to March are the months when production decreases due to a reduction in the herd, i.e. the animals have passed the "good" lactation period and consequently their production has decreased.

The figure below shows an analysis between the cost per liter of milk and the gross value received by the owner.

Figura 10 - Graph of the analysis between the cost per liter of milk and the gross value at market prices received by the owner

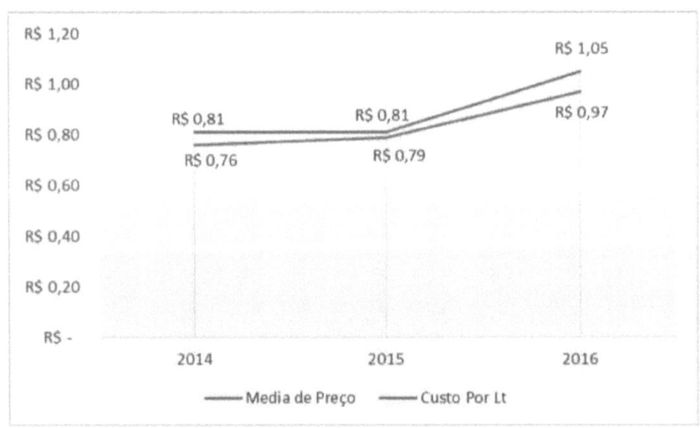

Source: Corso, Kalkmann and Ruppenthal 2017.

This graph shows the average cost per liter of milk in each year of the period studied, as well as the average prices charged by the owner. It can be seen that the average price between 2014 and 2015 remained the same, not discounting the inflation of the period, which in 2015 averaged 10.67% according to the IPCA (Consumer Price Index), meaning that the real gain in 2015 lost 10.67% of the purchasing power of 2014, meaning that the average real value received was only R$0.72 per liter of milk compared to 2014. However, with the increase in the average received in 2016, the inflation rate for the period should also be taken into account, which was 6.29 %, meaning that the real gain for producers was R$ 0.98 per liter of milk.

However, it can be seen that the cost of production has assiduously kept pace with the average sale price, i.e. in 2014 the "profit" was only R$0.06 per liter of milk, in 2015 this surplus was only R$0.02 per liter, but in 2016 there was a greater surplus in relation to the cost and average market price of only R$0.08 per liter of milk.

The owner uses a total of 5 hectares for his dairy farming activities, of which one part is where the handling facilities are built, another part is used for the farmyard, i.e. where the animals stay when they are not grazing, where the animals stay when they are not grazing, and the rest is used for pasture, where the rotational grazing system is used, i.e. the animals are managed in such a way that the areas are divided into paddocks that are subjected to alternating periods of grazing and resting. A major advantage is that you have greater control over the pasture, so you can define how long the plants will be grazed, which tends to be more uniform and feed more efficiently.

4.3.2 Mapping Maize Production

The property has been in the business for more than 40 years, and it began due to the need for animal feed on the property itself, where it produced only a small amount for its own consumption. Over the years, the area under cultivation has increased and the product has also begun to be marketed. Before, it was only used for consumption on the property to make feed with corn grain and silage was also produced as a whole plant. However, seeing the need for more income for the property, and analyzing the market, they saw an opportunity to increase the area produced to also market the grain.

In the period under study, the property had 10 hectares available in the 13/14 and 14/15 harvests, with an increase of 1 hectare in the 15/16 harvest. However, adding up the production and sales under study, counting the harvest and the off-season during the period, and the quantities used for the property's own consumption in dairy farming.

Figure 11 shows an analysis of the quantity produced, sold and consumed in each harvest period studied, in 60 kg bags.

Figura 11 - Graph comparing the quantity produced, the quantity consumed and the quantity sold of the corn crop

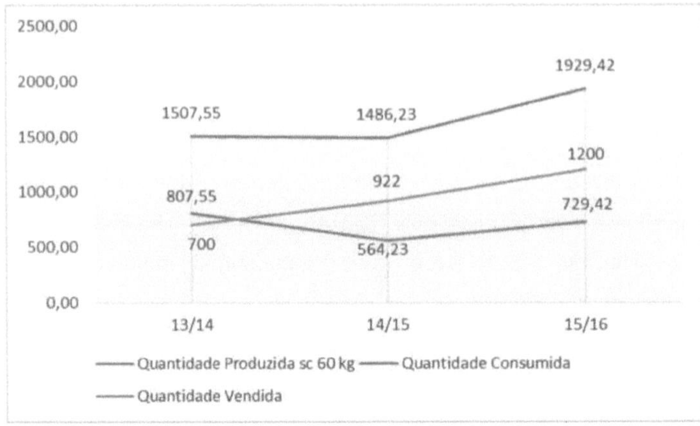

Source: Corso, Kalkmann and Ruppenthal 2017.

Figure 12 shows a stabilization in production between the 13/14 and 14/15 harvests. In the 13/14 harvest, consumption of production for dairy farming was higher than the quantity sold, which was not the case in the following years, when the quantity sold was higher than the quantity consumed. In the 15/16 harvest there was an increase of around 23% in production, due to the increase of 1 hectare in the area planted, and due to an excellent year of production in which the average production per hectare was 175.4 bags, compared to the 13/14 and 14/15 harvests which had an average production per hectare of 150.7 Sc and 148.6 Sc respectively.

Figura 12 - Graph of Gross Turnover, Total Marketed and Total Sold at market prices for the corn crop

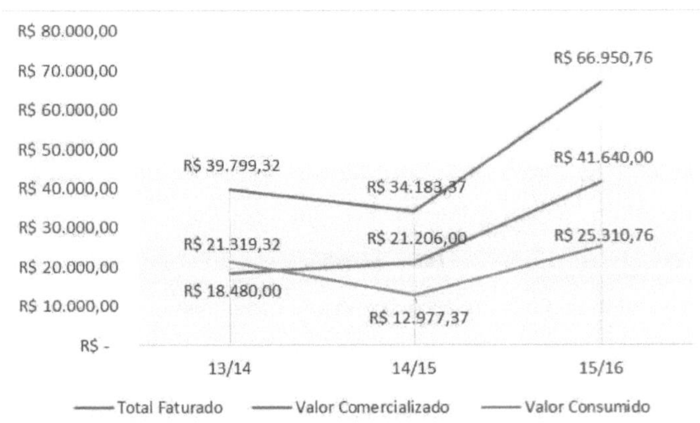

Source: Corso, Kalkmann and Ruppenthal 2017.

This graph shows in figures the gross invoiced amount, i.e. how much was invoiced from sales and internal consumption. It can be seen that from the 13/14 harvest to the 14/15 harvest, even though production was practically the same, there was a decrease of approximately 14% in the total amount invoiced. This is due to the fact that the average value marketed was also lower, so much so that in the 13/14 harvest a bag of the product was marketed at R$26.40 and in the 14/15 harvest at R$23.00, i.e. a drop in the average value of 12.87%.

Comparing previous years with the 15/16 harvest, there was an increase of more than 50% in the total gross revenue of the property, taking into account what was said above, plus the factor of the average value sold, which increased by 34% nominally, in real terms, compared to the 13/14 harvest, there was a decrease of 22.17%, even though there was an increase in the value of the bag sold, i.e. the average value per bag sold reached a level of R$34.70, which raised the percentage increase in the property's gross revenue to 25.90% in real market terms.

Figura 13 - Graph of the analysis between cost, turnover and profit at market prices per hc in corn cultivation

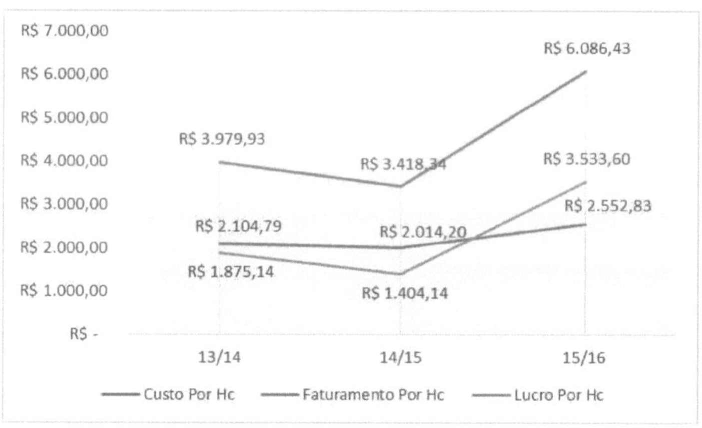

Source: Corso, Kalkmann and Ruppenthal, 2017.

Figure 13 shows the gross turnover per hectare, as well as the cost and profit. The average cost per hectare has not varied much over the years, due to the early purchase of inputs, thus guaranteeing a better profit margin per hectare. It can be seen that in the 14/15 harvest there was a drop in turnover and profit per hectare, as mentioned above, due to the drop in the average value invoiced, as well as in the 15/16 harvest, when turnover rose due to the increase in market prices, where profit became higher than cost, a fact that had not occurred in previous harvests. In the 15/16 harvest, the property made a gross profit of R$3,533.60, i.e. a 53% increase in gross profit compared to the 13/14 harvest when the research began, this increase being due to data already mentioned earlier in the text.

In terms of real market values, we have the year 13/14 as a base, where in the 14/15 harvest there was a 33.41% drop in its market value and in the 15/16 harvest this drop was 22.17%. In real terms, in the 15/16 harvest there was a 48.40% loss in value, i.e. at nominal prices the market price was R$34.70, but the real gain was only R$17.90, i.e. a huge difference when the values are taken at real market prices. Turnover fell by 34.34% in the 14/15 harvest compared to 13/14, and in the 15/16 harvest there was a nominal increase of 68.22% in gross turnover, but in reality there was only an increase of 1.06%, i.e. the gross amount invoiced was R$66,950.76, but in reality it was only R$34,546.59.

4.3.3 Mapping soybean production

The property has been working with this crop segment for over 40 years, as well as corn production, but this was always the main crop on the property until the period when milk began to play a bigger role on the property, thus causing the producer to reduce the area of soybeans planted, in order to use it more for dairy cattle management, as well as to reduce the area planted due to the increase in corn production, since it was also used for the property's internal consumption. This data is from a previous period than the data studied. The property had an available area of 7 hectares in the 13/14 and 14/15 harvests, with an increase of 3 hectares in production for the 15/16 harvest, due to the planting of the off-season crop, which in the case of this study is the production of the harvest as a whole.

Figure 14 shows the quantity produced and the average production per hectare.

Figure 14 - Graph of quantity produced and average production

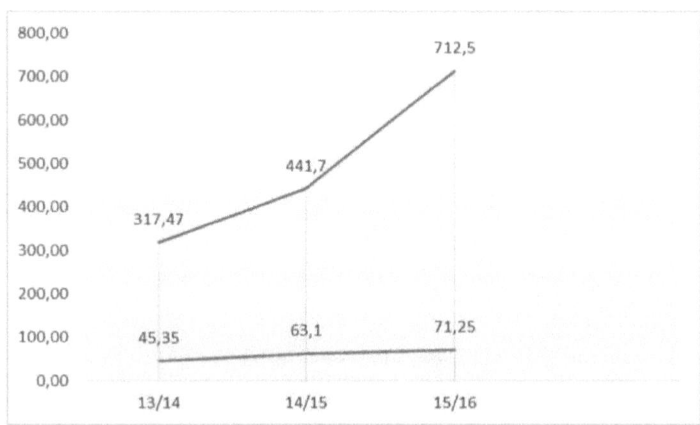

Source: Corso, Kalkmann and Ruppenthal, 2017.

The graph shows the evolution of the quantity produced as well as the evolution of productivity per hectare of the crop, i.e. from the 13/14 to the 14/15 harvest there was an increase in production of around 125 bags more with the same area of 7 hectares planted, also because productivity increased by 28%, directly impacting on the total produced, now, compared to the 15/16 harvest, there has been a 38% increase in total gross production, but this is also due to the 43% increase in the area planted, and productivity increasing by around 12%, directly impacting on production, whose year was excellent for this increase in production.

Based on gross revenue from soybean production, we can see the following figures, as shown in Figure 15.

Figura 15 - Graph of gross sales at market prices in soybean cultivation

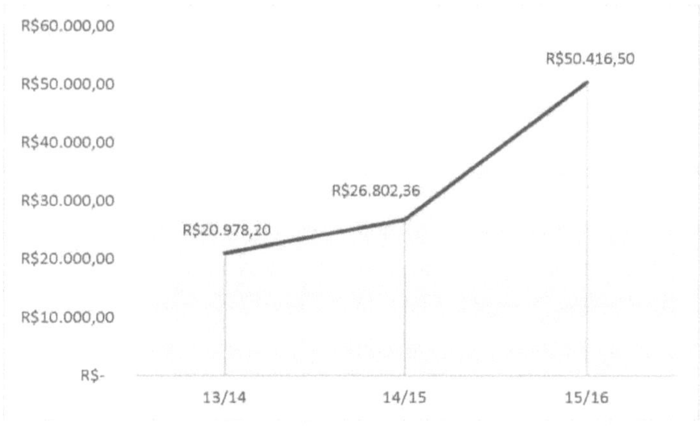

Source: Corso, Kalkmann and Ruppenthal 2017.

This data shows the total gross revenue per year/harvest for the activity, as well as the evolution of revenue, which has had a major impact on the property. Comparing the 15/16 harvest to the 13/14 harvest, there has been an increase of R$ 29. 433.30 in the property's revenue.438.30 increase in the property's turnover, i.e. a nominal increase of around 140 % in total, but the real increase, deflating the period, was only 42.88%, but all this taking into account the increase in the area planted and, nevertheless, the increase in the value of the *commodity sold on* the market, which in the case studied in the year/crop of 13/14 the sack of soya was sold for an average of R$ 66,08 and in the 15/16 harvest its average price rose to R$70.76, an increase of just 6.61%, which in real terms, with an average accumulated inflation of 23.37% according to the IPCA over the period, shows a loss in the purchasing power of the activity.

All of this has a direct impact on the activity's results, as shown in figure 16, which compares the activity's costs, gross revenue and profit per hectare.

Figura 16 - Graph comparing turnover, cost and profit at market prices for soybean cultivation by hc

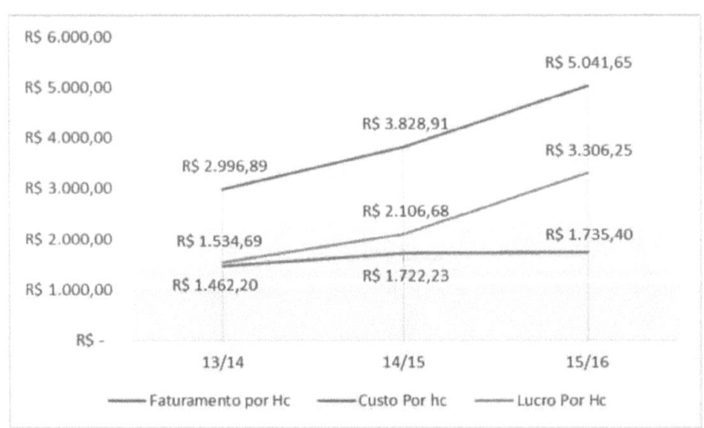

Source: Corso, Kalkmann and Ruppenthal, 2017.

This graph shows a comparison between the cost and profit of the activity, looking at a timeline between the harvests studied, it can be seen that the cost of production per hectare over the harvests suffers few changes, that is, it has a small increase of 18% less than the increase in inflation in the period, as described above, what has the greatest impact is the profitability per hectare, because from the 14/15 harvest to the 13/14 harvest there is an increase of 37,27% increase in profit and the 15/16 harvest has a 56.94%

increase in profitability, with a major positive impact on the property's cash flow, but this increase in profitability comes from a number of factors, such as the correct purchase of inputs at a lower cost, favorable weather that is suitable for production, and the value marketed, which also has a major impact on profitability.

4.4 ANALYSIS OF THE PROPERTY'S OPPORTUNITY COSTS

An analysis of the cost of rural property requires a lot of attention, as there are many factors that can be discussed and understood, as every activity has its own particularities, so everything requires a lot of planning and analysis in order to make decisions. The author then uses the *Microsoft Excel* solver tool to analyze the data obtained from the property and perform a mathematical analysis of the activities carried out.

The mathematical model was based on the following data: where x_1 means milk production, x_2 means corn production and x_3 soy production. The data was based on the availability offered by the property, with values, in the case of milk and corn production, segmented every 1 hectare, the costs so that the availability given in total value of the production area of the property, in order to have comparative data for the occupation of the entire cultivated area. The objective function for maximizing profits is given by the following expression: Max Z: $177.86X1 + 2{,}211.51X2 + 2{,}315.84X3$.

Table 1 - Mathematical model constraints

RESTRICTION	Milk		Corn		Soy		AVAILABLE	UN	
SC/hc fertilizer	0	+	5,8	+	6,67	≤	154	Sc	(1)
Urea SC/hc	0	+	3,87	+	0	≤	110	Sc	(2)
SC/hc Soybean Seed	0	+	0	+	1,25	≤	0	Sc	(3)
corn seed SC/hc	0	+	1	+	0	≤	22	Sc	(4)
pesticides R$/Hc	0	+	294,33	+	290,22	≤	6600	R$	(5)
Planting R$/hc	0	+	106,67	+	106,67	≤	2640	R$	(6)
Harvest sc	0	+	57,19	+	22,075	≤	1320	Sc	(7)
Feed and Bran R$/Month	3972,03	+	0	+	0	≤	4500	R$	(8)
Labor R$/Month	1216,67	+	0	+	0	≤	1300	R$	(9)
Pastures R$/Month	191,44	+	0	+	0	≤	4400	R$	(10)
Number of hectares	1	+	1	+	1	≤	22	Hc	(11)
No negativity	X1		X2		X3	≥	0		(12)

Source: Corso, Kalkmann and Ruppenthal, 2017.

Table 1 shows the constraint equations of the mathematical model. The equations from 1 to 7 are the inputs and pesticides used on each hectare of land to grow the two crops, which must be less than or equal to the availability of existing inputs and pesticides, so that the availability shown would be a total of the area, i.e. the availability of 22 hectares for planting. It can be seen that in some cases the consumption of a given input is zero, meaning that the agricultural input is not used for that crop. Observing that constraint 1, which is fertilizer in bags, is not used in dairy production, but in this activity it is linked to constraint 10 with a cost embedded in the value of the pasture. Restriction 2 is only used on corn crops because it increases the vigor and growth of the plant, and it is not necessary to apply it to soybean crops.

It should also be noted that restrictions 3 and 4 refer to bags of seeds used for planting, while restriction 5 is based on the cost in value of the pesticides used for soybean and corn cultivars, i.e. any and all pesticides used from soil preparation with herbicides to pest control with fungicides. Restrictions 6 and 7 are based on planting values and the percentage of bags harvested, respectively, where planting is fixed at a value per hectare, and harvesting at a percentage of the gross value produced (planting and harvesting are the costs of the labor used to carry out the activities). The data analyzed is based on the average of the years studied to compile the model.

Constraints 8 to 10 refer exclusively to dairy farming, where they include the average costs used for feed and meal, labor, and the cost of pasture for animal feed, and constraint 11, which is based on the number of hectares available on the property for the production of the activities. Equation 12, the last one, refers to the fact that the variables need to be greater than or equal to zero, i.e. non-negativity indicates that the variables cannot be less than zero.

4.4.1 Solving the mathematical model

To solve the mathematical model, the three decision variables, the objective function and the 11 constraints were entered into *Microsoft Excel.* Table 3 below shows the modeling of the original problem in *Excel.*

Chart 3- Modeling the original problem in Microsoft Excel software

	Milk $(x)_1$	Maize $(x)_2$	Soy (x_3)	
Maximize Z=	177,86	2211,51	2315,84	Profit
Variable cells	0	0	0	R$ -

				Used	Available	Accessibility	Unit
SC/hc fertilizer	0	5,8	6,67	-	≤	154	Sc
Urea SC/hc	0	3,87	0	-	≤	110	Sc
SC/hc Soybean Seed	0	0	1,25	-	≤	28	Sc
corn seed SC/hc	0	1	0	-	≤	22	Sc
pesticides R$/Hc	0	294,33	290,22	-	≤	6600	R$
Planting R$/hc	0	106,67	106,67	-	≤	2640	R$
Harvest sc	0	57,19	22,08	-	≤	1320	Sc
Feed and Bran R$/Month	3972,03	0	0	-	≤	4500	R$
Labor R$/Month	1216,67	0	0	-	≤	1300	R$
Pastures R$/Month	191,44	0	0	-	≤	4400	R$
Number of hectares	1	1	1	-	≤	22	Hc

Source: Corso, Kalkmann and Ruppenthal, 2017.

As can be seen in the table, the restrictions are based on the quantity of products used in the activities. As highlighted in the figure, restrictions 1,2,3,4 and 7 are shown in the unit of bags of the product described per hectare, while restrictions 5,6,8,9 and 10 are expressed in values, i.e. the value of the cost per hectare, the quantity used or the cost value are located on the left of the table, while the availability of the property is designated on the right. The less than or equal to sign shows that the equations must be less than or equal to the amount of resources available. The objective function is shown at the top with the three decision variables, xi (milk), x_2 (corn) and x_3 (soy).

4.4.2 Analysis of the mathematical model for decision making

As already mentioned in the methodology, the ideal mathematical models for the property under study are analyzed below.

Decision theory is a set of knowledge and analytical techniques related to different degrees of formality, developed to help the decision maker process choices between a scenario of alternatives, taking into account the possible consequences of the decision, with the aim of selecting alternatives that represent a preferred set of consequences (OLIVEIRA, *et al,* 2007).

An analysis of a property's results requires a great deal of care and rigor in the data analyzed, as it seeks to identify the best alternatives to focus on and redirect the property's investments. An analysis that lacks data or is even incomplete can be detrimental to the final result of the analysis and so decision-making can be unsuccessful.

Three sensitivity reports were put together, where calculations were made for profitability by producing the total area with corn, another with soybeans, and another with joint production, i.e. half the area used for corn and half used for soybeans, since this would be one of the most suitable through crop rotation, pointing out that the costs and values practiced are at present values and these may vary, either up or down in the coming years. From the data obtained, the following results were found.

Table 4 - Soybean production model

	Milk $(x)_1$	Maize $(x)_2$	Soy $(x)_3$				
Maximize Z=	177,86	2211,51	2315,84	Profit			
Variable cells	0	0	22	50948,48			
				Used	Available	Accessibility	Unit
SC/hc fertilizer	0	5,8	6,67	146,74	≤	154	Sc
Urea SC/hc	0	3,87	0	0	≤	110	Sc
SC/hc Soybean Seed	0	0	1,25	27,5	≤	28	Sc
corn seed SC/hc	0	1	0	0	≤	22	Sc
pesticides R$/Hc	0	294,33	290,22	6384,84	≤	6600	R$
Planting R$/hc	0	106,67	106,67	2346,74	≤	2640	R$
Harvest sc	0	57,19	22,08	485,65	≤	1320	Sc
Feed and Bran R$/Month	3972,03	0	0	0	≤	4500	R$
Labor R$/Month	1216,67	0	0	0	≤	1300	R$
Pastures R$/Month	191,44	0	0	0	≤	4400	R$
Number of hectares	1	1	1	22	≤	22	Hc

Source: Corso, Kalkmann and Ruppenthal 2017.

As shown by the data obtained from the mathematical model, the activity that would maximize the profits to be developed by the property would be soybean cultivation, which brings a profitability of R$ 50.948.48 per harvest, producing the 22 hectares available on the property and using the resources that the property has to offer, which is higher than producing corn alone, which would have a profitability of R$ 48,653.22, and dairy farming would be completely unviable for the property, i.e. there would have to be a reduction of at least R$ 2137.98 in the cost of dairy farming in order to start making a profit.

Table 5 - Joint production model

Milk (x_1) Corn (x_2) Soy (x_3)						
Maximize Z=	177,86	2211,51	2315,84	Profit		
Variable cells	0	10,8	11,2	R$ 49.821,72		
				Used	Availability	Unit
SC/hc fertilizer	0	5,8	6,67	137,344	≤	154 Sc
Urea SC/hc	0	3,87	0	41,796	≤	110 Sc
SC/hc Soybean Seed	0	0	1,25	14	≤	14 Sc
Maize seed SC/hc	0	1	0	10,8	≤	11 Sc
Defensives R$/Hc	0	294,33	290,22	6429,228	≤	6600 R$
Planting R$/hc	0	106,67	106,67	2346,74	≤	2640 R$
Harvest sc	0	57,19	22,08	864,892	≤	1320 Sc
Feed and Bran R$/Month	3972,03	0	0	0	≤	4500 R$
Labor R$/Month	1216,67	0	0	0	≤	1300 R$
Pastures R$/Month	191,44	0	0	0	≤	4400 R$
Number of hectares	1	1	1	22	≤	22 Hc

Source: Corso, Kalkmann and Ruppenthal 2017.

Even though the mathematical model presents only soybean production as ideal, because it maximizes profit, we suggest joint production, because as described above, there are other factors that interfere with production. Thus, the land depends on special management in order not to deplete its productivity, and crop rotation is one of the ideal aspects to strengthen the land in production.

Crop rotation is understood to be the regular and orderly alternation in the cultivation of different plant species in temporal sequence in a given area (MUZILLI et al., 1983). The use of green manures and crop rotation in conservation management has proved indispensable since the beginning of research into the no-till system. Crop rotation consists of alternating plant species, within the same agricultural period, throughout the years of cultivation, in the same agricultural area. Cover crops serve to form mulch on the soil surface, culminating in a reduction in spending on nitrogen fertilizers and herbicides

(AITA et al., 1994).

Crop rotations alternated annually bring numerous advantages for production, as well as for the physical improvement of the soil, since they control weeds, diseases and pests, replenish organic matter and protect against the action of climatic agents on the soil. It is also necessary to consider that it is not enough just to follow a sequence of crops, but the farmer must involve technology such as techniques for good plant cultivation and disease control, to strengthen the soil, making it more productive (IEPEC, 2017).

The limitations on the ideal choice of planting are noticeable, given the different variables that alter the cost composition of the recommended activities. In the case of milk, production is unfeasible because its costs are very high and its margin of return is minimal, and may even be negative in some periods under certain market circumstances, *coeteris paribus with* the other variables.

Corn production is more competitive than dairy farming, given that the margin of return is higher per unit of measurement, although it has higher costs when compared to soybean cultivation, taking into account the analysis of the other constant variables, including the wear and tear on the land factor, which is essential for production.

Due to its value on the market, its productivity and its reduced cost compared to the other variables in the study, soybean production is closer to a more rational model in terms of the use of resources applied to economic activity, with a view to maximizing profits. Given the limitations of the mathematical model, it can be seen that planting this crop is the most appropriate and profitable.

In view of this analysis, we suggest alternating the planting of soybeans and corn on the property, even if the profit decreases by a not very significant amount, around 2.2%, but over time this value may decrease due to the impoverishment of the soil, resulting in lower productivity, because even so with the alternation of activities the property would maximize its profits according to the mathematical model by R$ 49,821.72 annually and the profit from soybean cultivation would be R$ 50,948.48.

It should be pointed out that all the cost and revenue figures were based on current market values, but these values can change over time, as they depend on various factors that affect the final profit, such as the price charged in the market, the cost of inputs, the climate that affects production, among other factors that can affect the final profit.

The sensitivity report is presented, which provides important economic analysis for decision-making.

Table 6 - Sensitivity analysis

Variable cells						
Cell	**Name**	**Final Value**	**Low Cost**	**Objective Coefficient**	**Allowed Increase**	**Allowed Reduce**
B3	Milk $(x)_1$	0	-2033,65	177,86	2033,65	1E+30
C3	Maize $(x)_2$	10,8	0	2211,51	104,33	2033,65
D3	Soy (x_3)	11,2	0	2315,84	1E+30	104,33
Restrictions						
Cell	**Name**	**Final Value**	**Shade Price**	**Lateral H.R. Restriction**	**Allowed Increase**	**Allowed Reduce**
E5	SC/hc fertilizer used	137,344	0	154	1E+30	16,656
E6	Urea SC/hc Used	41,796	0	110	1E+30	68,204
E7	Soybean Seed SC/hc Used	14	83,464	14	13,5	0,25
E8	corn seed SC/hc Used	10,8	0	11	1E+30	0,2
E9	pesticides R$/Hc Used	6429,228	0	6600	1E+30	170,772
E10	Planting R$/hc Used	2346,74	0	2640	1E+30	293,26
E11	Harvest sc Used	864,892	0	1320	1E+30	455,108
E12	Feed and Bran R$/Month Used	0	0	4500	1E+30	4500
E13	Labor R$/Month Used	0	0	1300	1E+30	1300
E14	Pastures R$/Month Used	0	0	4400	1E+30	4400
E15	Number of hectares used	22	2211,51	22	0,2	10,8

Source: Corso, Kalkmann and Ruppenthal 2017.

As can be seen from the sensitivity report, in order for dairy farming to be viable, costs must be reduced by at least R$2,033.65 per month so that, from this point on, the property can generate profitability. Under the current conditions of the property, it is suggested that a total of 10.8 hectares of corn and 11.2 hectares of soybeans be produced, in order to maximize profits.

The focus of this study is on the theory of the firm within the neoclassical view of profit maximization, and it did not take into account other factors that determine alternative views of economic theories, such as the theories of the Austrian school, the neo-Shumpeterians, institutionalists, among other currents of evolutionary thought. It was limited in this way in view of the empirical evidence of the *mainstay of* neoclassical economic thinking on the maximization of a company's profits.

The suggestions about the ideal mathematical model do not significantly alter the precepts defended at the top of this study, but are merely a suggested temporal decision, in view of the scenario and significance of the variables presented in the study during this period of analysis. Therefore, it can be emphasized that the study is open to further and alternative analyses of the different currents of economic thought, which does not detract from the present, but limits it in the scope of its analysis.

5 FINAL CONSIDERATIONS

An economic analysis for small farms is useful for making decisions, but it requires a very strong foundation when high values are involved. Otherwise, you run the risk of making the wrong investments, or even failing to allocate a resource that would bring a greater return to the property. Planning for small farms today is of the utmost importance. Often, more concrete planning doesn't take place on small properties, as producers generally have a low income and make their living from the small productive area available on their property. As a rule, these producers don't have a good basis and the knowledge to plan their production or improve their results.

Price fluctuations bring great instability and uncertainty to rural producers in terms of their gross income, as their costs generally remain at the same level, but profitability fluctuates according to the market and the price charged varies greatly. This makes it very important for producers to maintain productivity or always aim to reduce their production costs, always bearing in mind the uncertainty about the sales prices charged on the market. Farmers' lack of awareness of real variables has a major impact, because as mentioned throughout the paper, there is a big difference between nominal values and real values, i.e. inflation has a major impact on turnover and final profitability, often going unnoticed by the producer who has a "false vision" of his business, thinking that he is working at a profit, but bringing it to real corrected values, he may be losing his purchasing power compared to previous years.

The main objective of this study was to study and characterize the tools used to manage economic activity on a rural property. In this sense, it was initially necessary to carry out a bibliographical study on the subject, and then apply the concepts to the property under study, where data was collected through interviews for analysis and a better understanding of the property's current situation. Based on this, it was possible to achieve the objectives of this study.

The first objective was to "Map the factors of production and detail the economic activities currently carried out on the property under study", where the data obtained from the property made it possible to identify the current situation by collecting data such as: activities carried out, business methods, availability to carry out the activities, market, among other variables.

Based on the mapping and detailing of economic activities, the objective of "carrying out a survey of expenses and income on the property studied" was also met, where all the income and costs involved in the property were calculated in spreadsheets.

In order to achieve the objective of "Identifying the opportunity costs between alternative activities", a linear programming mathematical model was developed, which showed the optimum production of the activities that maximize profit. Although the mathematical model gave the optimum result for the production of just one crop, it is suggested that two crops be grown simultaneously, so that there is crop rotation, because as mentioned in the text there are several factors that influence production.

Also based on the mathematical model, the objective of evaluating the activities that maximize and enhance the economic results of the property studied can be met, presenting concrete data that can be used to make decisions for the producer.

In response to the research problem presented in the study of how to optimize production with the resources available and control the activities carried out through an economic analysis on a small rural property in the municipality of Tucunduva, it can be said that an activity, even if it has a certain turnover, does not mean that it is profitable. This can be seen in the analysis of dairy farming, which, even though it has a reasonably good turnover, is not profitable for the specific property under study.

In this sense, the mathematical model can show the most profitable activities with the resources available on the property, and is considered an appropriate tool for management and economic analysis, showing the reality of a property through concrete numbers, serving as a basis for decision-making.

Given the farmer's desire to remain in dairy farming, we suggest that future work should include ways of reducing the costs involved in the activity, so that it can become viable and profitable for the property. It is also suggested that this study be validated annually because its variables fluctuate over time.

6 REFERENCES

AITA, C. et al. Winter species as a source of nitrogen for corn in the minimum tillage system and beans in no-tillage. **Revista Brasileira da Ciência do Solo**, v. 18: n. 1, p. 101-108, 1994

ARAÙJO, Massilon J. **Fundamentos de Agronegócios**. 2.ed. Sao Paulo: Atlas, 2005.

BACHA, C.J.C. **Economia e Politica Agricola no Brasil**. Sao Paulo: Atlas, 2004. Chapter 3, item on sugar-alcohol policy. 226 p.

BACHA, C.J.C. **Entendendo a Economia Brasileira**. Campinas: Atomo, 2007.

BARROS, A.J.S e LEHFELD, N.A.S. **Fundamentos de Metodologia**: Um Guia Para Iniciaçao Cientifica. 2 ed. Sao Paulo: Makron Books, 2000.

BILAS, Richard A. **Teoria microeconômica**. 8. ed. Rio de Janeiro: Forense- Universitària, 1980.

BRANDT, GILIANE TROST et al, **Family Succession in** Agribusiness **Companies** - Master's Degree - Federal University of Rio Grande do Sul, School of Administration, Postgraduate Program in Administration, Porto Alegre, BR-RS, 2015. Available at:

<http://www.lume.ufrgs.br/handle/10183/111797> Accessed on: September 16, 2016

BRAUN, Adeli Beatriz. *Et al.* **Analysis of a Family Business:** A Study of the Management of a Farm Located in the Northwest of Rio Grande do Sul. VIII Symposium on Excellence in Management and Technology. Resende, 2011.

BUAINAIN, A. M.; ROMEIRO, A. R.; GUANZIROLI, C. **Family Farming and the New Rural World.** Sociologias, Porto Alegre, ano 5, n. 10, jul/dez 2003, p. 312-347

CALDERELLI, Antonio. **Enciclopédia contàbil e comercial brasileira.** 30. ed. Sao Paulo: CETEC, 2003.

CECCONELLO, Antonio Renato; AJZENTAL, Alberto. **The construction of a business plan**: methodological path for: characterizing the opportunity, structuring the conceptual project, understanding the context, defining the business, developing the strategy, sizing the operations, projecting results, feasibility analysis. Sao Paulo: Saraiva, 2008.

CoDAF. **The importance of family farming**. Available at:

<http://codaf.tupa.unesp.br/informaçôes/a-importancia-da-agricultura-familiar>.

Accessed on: November 18, 2016.

CONTINI, Elisio. **Dynamism of Brazilian agribusiness**. Available at: <http://www.agronline.com.br/artigos/artigo>. Accessed on: November 12, 2016.

CREPALDI, Silvio Aparecido. **Rural accounting: a decision-making approach**. 4. ed. revised, updated and expanded. Sao Paulo: Atlas, 2006.

DANTAS, A. *et. al.* **Company, industry and markets**. In: KUPFER, D.; HASENCLEVER, L. (Org.). Economia industrial: fundamentos teóricos e pràticas no Brasil. Rio de Janeiro: Campus. p. 23-41, 2002.

GERSICK, Kelin; DAVIS, John A.; HAMPTON, Marion M.; LANSBERG, Ivan. **From generation to generation:** life cycles of family businesses. Rio de Janeiro: Elsevier, 2006.

GIL, Antônio Carlos. **How to design research projects**. 4 ed. Sao Paulo: Atlas, 2002.

GIL, Antonio Carlos. **Methodology of higher education**. 4 ed. Sao Paulo: Atlas, 2005.

GOMES, Aguinaldo Rocha. **Rural accounting & family farming.** Rondonópolis: A. R. Gomes, 2002.

GUILHOTO, J. J. M *et al.* **GDP of Family Farming**: Brazil - States. NEAD Studies. Ministry of Agrarian Development (MDA). Brasilia, 2000.

GUILHOTO, J. J. M. **Brazilian agribusiness between 1959 and 1995**: economic dimension, structural change and trends. P 3-32. In: Brazilian agribusiness at the end of the 20th century. Passo Fundo-RS: UPF, 2000. 2 v. 337 pag.

HAMBRICK, D. **Operationalizing the concept of business strategy in research**. Academy of Management Review, vol. 5, n. 4, p. 567-575, 1980.

IBGE - Brazilian Institute of Geography and Statistics. **Agricultural census**. Rio de Janeiro, p.1-267, 2006. Available at: <http://biblioteca.ibge.gov.br/visualizacao/periodicos/50/agro_2006_agricultura_familiar.pdf>. Accessed on November 12, 2016.

IEPEC, **The importance of applying crop rotation.** Sao Paulo, 2017, available at: <http://iepec.com/importancia-de-aplicar-rotacao-de-culturas/->, Accessed on May 4, 2017.

INCRA, National Institute for Colonization and Agrarian Reform. **Table of municipal fiscal modules**. Available at: <http://www.incra.gov.br/tabela- modulofiscal>. Accessed on: April 14, 2017.

LAKATOS, Eva Maria Marconi; ANDRADE, Marina de. **Metodologia do trabalho cientifico.** 5 ed. Sao Paulo: Atlas, 2001.

LEFTWICH, Richard H. **Introduction to microeconomics**. New York, Holt, Renehart and Winston, 1970

LIPSEY, R.G. & STEINER, P.O. **Economics.** 2nd ed. New York, Harper & Row, 1969.

MARION. José Carlos. **Corporate Accounting**. 10. ed. Sao Paulo: Atlas, 2003.

MARTINS, Gilberto de Andrade; THEÓPHILO, Carlos Renato. **Metodologia da investigaçâo cientifica para ciências sociais aplicadas**. Sao Paulo: Atlas, 2002.

MAURO, Carlos Alberto. **Transfer Pricing Based on Opportunity Cost**: An Instrument for Promoting Business Efficiency. Master's dissertation. Sao Paulo: FEA-USP, 1991.

MENDES, J. T. G.; JUNIOR, J. B. P; **Agronegócio:** uma abordagem econômica. Sao Paulo: Pearson Prentice Hall, 2007.

MEYERS, Albert L. **Modern economics**: elements and problems. New York, Prentice-Hall, 1942.

MILLER, Roger LeRoy. **Microeconomics**: theory, issues and applications. Sao Paulo: McGrawHill, 1981.

MUZILLI, O. et al. Nitrogen fertilization in maize in Paranà. III. Influence of soil recovery with winter green manure on responses to nitrogen fertilization. **Revista Brasileira da Ciência do Solo**, Campinas, v. 18, n. 1, p. 2327, 1983.

OLIVEIRA, Djalma de Pinho Rebouças de. **Sistemas de Informaçôes Gerenciais**: Estratégicas Tàticas Operacionais. 12 ed. Sao Paulo: Editora Atlas, 2008.

OLIVEIRA, L. M. et al. **Information and the rural producer's decision-making process.** SOBER, 2007. Available at: <http://www.sober.org.br/palestra/6/261.pdf>>. Accessed on: April 19, 2017.

PADOVANE, Luiz Clóvis. **Introduction to Accounting, with an approach for non-accountants.** Sao Paulo: Ed. Thomson, 2006.

PASSOS, Édio; *et al.* **Familia, Familia, negócios à parte:** como fortalecer laços e desatar nós na empresa familiar. 5.ed. Sao Paulo: Editora Gente, 2006.

PINAZZA, Luiz Antonio, ARAÙJO, Ney Bittencourt de. **Agriculture at the turn of the 20th century: the** vision of Agrobusiness-

PINDYCK, R. S.; RUBINFELD, D. L. **Microeconomics**. 5. ed. Sao Paulo: Prentice Hall, 2009.

PORTER, M. E. **Competitive strategy:** techniques for analyzing industries and competitors. New York : Free Press, 1980.

POSSAS, M. **Estrutura de mercado em oligopólio**. Sao Paulo: Hucitec, 1985.

PRAHALAD, C.K. HAMEL, G. **Competing for the Future**. Innovative strategies for taking control of your industry and creating the markets of tomorrow. Rio de Janeiro: Campus, 1995.

QUINN, James B. **The strategy process**. 3. ed. Porto Alegre: Bookman, 2001.

REIS, Alida T. S. *et al.* **Accounting applied to agribusiness**: Analysis of the company Rasip Agropastoril S.A. Available at <http://www.sinescontabil.com.br/monografias/artigos/CONTABILIDADE-APLICADA-ATO-AGRONEGOCIO.PDF> Accessed on: October 25, 2016.

RIOS, Terezinha Azerêdo. The presence of philosophy and ethics in the professional context. **Organicom Magazine**, Sao Paulo, v. 5, n. 8, 2008. Available at: <http://revistaorganicom.org.br/sistema/index.php/organicom/article/view/145/245>. Accessed on: Aug. 16, 2016.

ROSA, José Antônio; MARÓSTICA, Eduardo. **Business models: organization and management.** 1. ed. Sao Paulo, 2013.

RUFINO, J. L. dos S. **Origin and concept of agribusiness**. Informe Agropecuário, Belo Horizonte, v. 20, n. 199, p. 17-19, 1999.

SANTOS, Gilberto José; MARION, Jose Carlos; SEGATTI, Sonia. **Administraçao de custos na Agropecuâria.** 3. ed. Sao Paulo: Atlas, 2002.

SAULIT, Jandrei Luis. **Agronegócio em ascendência**. 2.ed. Rio de Janeiro: Atlas, 1997.

SCHUMPETER, Joseph Alois. **Theory of economic development**: an investigation into profits, capital, credit, interest and the business cycle. Translated by Maria Silvia Possas. 3. ed. Sao Paulo: Nova Cultural, 1988.

SETTE, R. de S. Administraçao **estratégica na empresa rural**. In: 3 Congresso Brasileiro de Administraçao Rural: Administraçao rural & agronegócio no 3 milênio, 1999, Anais, p. 51 - 63.

SLACK, N.; CHAMBERS, S.; JOHNSTON, R. **Administraçâo da Produçâo**. 2.ed. Sao Paulo: Atlas, 2002.

STIGLITZ, J. E., **Introduction to microeconomics**. Rio de Janeiro : Campus, 2003 - 7ª reprint

STOFFEL, Janete. **The influence of family farming on rural development in southern Brazil,** 2013. Available at: <http://repositorio.unisc.br/jspui/bitstream/11624/527/1/JaneteStoffel.pdf> Accessed on June 5, 2017.

TODESCO, Joao Carlos. **Family farming**: realities and prospects. Passo Fundo: EDIUPF, 1999.

VARIAN, Hal R. **Microeconomics**: basic principles. Rio de Janeiro: Campus, 1994.

VELOSO, Rui Fonsêca; FERNANDES, Fernando Borges; BARIONI, Luis Gustavo. **The importance of financial control in a management information system on a family farm.** In: IV Congresso brasileiro da sociedade brasileira de informàtica aplicada à agropecuària e à agroindùstria. Proceedings, 2003. Accessed on: September 28, 2016.

VICARI, Flavio Marques. **Business Strategy**: Brazilian Cases. 2.ed. Jundiai: Paco Editorial, 2013.

WERNER, René. **Family and business**: a path to success. Sao Paulo: Manole, 2004. The family business and Agribusiness. Gazeta de Ribeirao magazine, p. 2, August 2006.

WILDAUER, Egon Walter. **Business plan**: constituent elements and preparation process. Curitiba: Ipbex, 2010.

YIN, Robert K. **Case study**: planning and methods. 3. ed. Sao Paulo: Bookman, 2005.

Case study: planning and methods. 2. ed. Porto Alegre: Editora Bookmam, 2001.

APPENDIX A - QUESTIONNAIRE APPLIED TO THE OWNER

1- - Have you ever heard of a business plan? Have you had training

2- How are production costs managed?

3- - Do you use any technological artifacts? Spreadsheets and computer? Do you?

4- Is the correct amount of inputs used in the activities carried out on the property?

5- Is there any possibility of reducing costs? In your opinion, which method can be used?

6- Do you do price research when selling inputs? What is the cost of logistics?

7- What are the threats and opportunities of the activities carried out on the property?

8-Are you thinking of changing any of your activities? If yes, why?

9 How long have you been in business? What has improved since you started? If so, what has changed and why? If not, why hasn't it changed, and how difficult was it?

10- Has innovation/technology had an impact on the business?

12- Do you research prices in more than one business when buying inputs?

MIX
Papier aus verantwortungsvollen Quellen
Paper from responsible sources
FSC® C105338

Printed by Books on Demand GmbH, Norderstedt / Germany